Rigging Pocket Guide
A Reference for the Rigging Professional

ISBN 978-1-8-8872401-1 **Revised: July 2016**

Text and illustrations ©2007-2016 by Jerry Klinke

OSHA 1910.184 Whenever any sling is used the following **practices shall** be observed:

1. Slings that ar
2. Slings shall no makeshift de
3. Sling legs shall
4. Slings shall no in excess of their rated capacities.
5. Slings used in a basket hitch shall have the loads balanced to prevent slippage.
6. Slings shall be securely attached to their loads.
7. Slings shall be padded or protected from the sharp edges of their loads.
8. Suspended loads shall be kept clear of all obstructions.
9. All employees shall be kept clear of loads about the be lifted and of suspended loads.
10. Hands or fingers shall not be placed between the sling and its load while the sling is being tightened around the load.
11 Shock loading is prohibited.
12. A sling shall not be pulled from under a load when the load is resting on the sling.

Inspections - Each day before being used, the sling and all fastenings and attachments shall be inspected for damage or defects by a competent person designated by the employer. Additional inspections shall be performed during sling use where service conditions warrant. Damaged or defective slings shall be immediately removed from service.

This publication IS NOT intended to replace formal training related to crane and rigging operations.

This publication DOES NOT provide a comprehensive or exhaustive list of all the possible situations that are encountered with rigging and hoisting operations. This publication is intended only to assist the user by providing typical equipment capacities and general guidelines that may be useful with planning.

Always select and use rigging equipment by the rated capacity shown on the ID tag, equipment labels or as provided by the manufacturer.

Published by:
ACRA Enterprises, Inc.
"Rigging Training and Publications"
5950 Red Arrow Hwy
Stevensville, MI 49127
800-992-0689 or 269-429-6240
www.acratech.com

Created and Printed in the USA

The information contained in this publication was compiled from sources believed to be reliable. It should not be assumed that this guide covers all regulations or standards. The author and publisher cannot guarantee correctness or completeness and accepts no responsibility in the use or misuse of this information.

The entire contents of this guide are copyrighted and protected by U.S. laws and cannot be reproduced without written permission of the author or publisher.

Table of Contents

Basic Knots
(for use with tagline ROPE)

BOWLINE

A bowline is a very secure knot which won't slip, regardless of the load applied.

CLOVE HITCH

This is a popular knot for securing to posts, bars, and other round objects.

SHEET BEND

A useful knot for tying two ropes together, even if the rope sizes and materials differ greatly.

DOUBLE SHEET BEND

This knot provides greater security, especially in plastic rope. Its the same as the sheet bend but with an extra coil around the standing loop.

Basic Hitches

VERTICAL

eye & eye — Wire rope / Web Sling

Chain Sling

Web / Roundsling — *endless*

CHOKER

Working Load Limits shown on sling tags and other charts are based on the angle of choke to be 120 degrees or greater.

120°

BASKET

Working Load Limits shown on sling tags and other charts are based on both legs being vertical within 5 degrees.

5° 5°

90°

90°

Hitch Reductions

Capacity tags show the Working Load Limit (WLL) for each hitch based on the examples on the left.

VERTICAL 10,000 lbs WLL	CHOKER 7,500 lbs WLL	BASKET 20,000 lbs WLL

VERTICAL

100%

When angles, a smaller D/d ratios, or a tight choke is used, the capacity of the hitch is MUCH LESS than what is shown on the tag!

2 slings @ 30 degrees horizontal angle

120° 50% reduction

CHOKER

100%

≥120°

90° 87%

29° 49%

BASKET

100%

90°

Low horizontal angles have a great effect on the hitch capacity!

60° 87%

30° 50%

v5 r1-5/17

Important Information

Effective on July 8, 2011 - OSHA updated the standards on slings for general industry (§1910.184) and construction (§1926.251)

1926.251(c)(16) - Wire rope slings shall have permanently affixed, identification tags stating size, rated capacity for the type(s) of hitch(es) used and the angle upon which it is based, and the number of legs if more than one.

1926.251(f)(1) - Shackles require legible identification markings as prescribed by the manufacturer.

The capacity tables previously designated in the OSHA standards are now obsolete. Now ALL wire rope slings must be marked - no exceptions! OSHA prohibits the use of ANY sling that does not have a permanently affixed identification tag.

Sling identification MUST be done by the sling manufacturer

All Working Load Limits (WLL) are based upon the items being in new, unused condition.

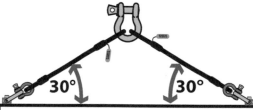

30°   30°

ASME B30.9-1.5.3 states "Horizontal sling angles less than 30 deg shall not be used except as recommended by the sling manufacturer or a qualified person."

EIPS - MS

IWRC

6x19 and 6x36 class

Extra Improved Plow Steel, Mechanical Splice

120° or greater

90°

SIZE	VERTICAL	CHOKER	BASKET
1/4"	0.65	0.48	1.3
5/16"	1	0.74	2
3/8"	1.4	1.1	2.9
7/16"	1.9	1.4	3.9
1/2"	2.5	1.9	5.1
9/16"	3.2	2.4	6.4
5/8"	3.9	2.9	7.8
3/4"	5.6	4.1	11
7/8"	7.6	5.6	15
1"	9.8	7.2	20
1-1/8"	12	9.1	24
1-1/4"	15	11	30
1-3/8"	18	13	36
1-1/2"	21	16	42
1-5/8"	24	18	49
1-3/4"	28	21	57
1-7/8"	32	24	64
2"	37	28	73

Based on generic data - Always use the sling tag to obtain the Working Load Limits (WLL)

Note: Rated loads based on a minimum D/d of 25:1

NOTE: The capacity tables are only intended to assist the user by providing typical sling data for job planning activities

v5 r1- 5/17

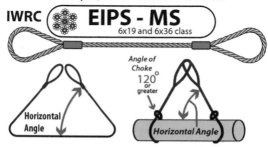

IWRC EIPS - MS
6x19 and 6x36 class

Angle of Choke
120° or greater

Horizontal Angle

Horizontal Angle

Basket			2-Chokers	
60°	**30°**	**SIZE**	**60°**	**30°**
1.1	0.65	**1/4"**	0.82	0.48
1.7	1	**5/16"**	1.3	0.74
2.5	1.4	**3/8"**	1.8	1.1
3.4	1.9	**7/16"**	2.5	1.4
4.4	2.5	**1/2"**	3.2	1.9
5.5	3.2	**9/16"**	4.1	2.4
6.8	3.9	**5/8"**	5	2.9
9.7	5.6	**3/4"**	7.1	4.1
13	7.6	**7/8"**	9.7	5.6
17	9.8	**1"**	13	7.2
21	12	**1-1/8"**	16	9.1
26	15	**1-1/4"**	19	11
31	18	**1-3/8"**	23	13
37	21	**1-1/2"**	28	16
42	24	**1-5/8"**	32	18
49	28	**1-3/4"**	37	21
56	32	**1-7/8"**	42	24
63	37	**2"**	48	28

Note: Rated loads based on a minimum D/d of 25:1

The values shown are listed in US Tons and are based on generic data
Always consult the sling manufacturer for the exact Working Load Limits

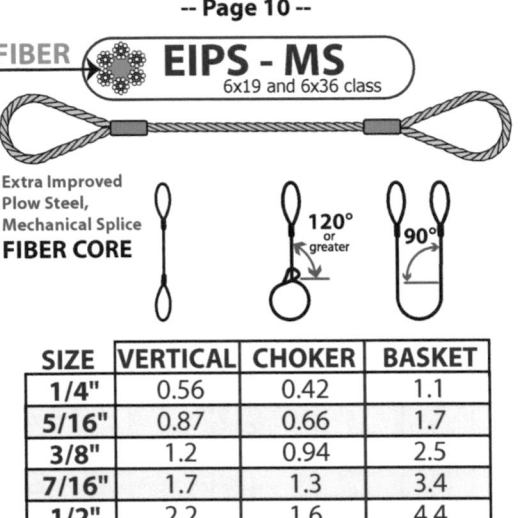

FIBER

EIPS - MS
6x19 and 6x36 class

Extra Improved
Plow Steel,
Mechanical Splice
FIBER CORE

120° or greater

90°

SIZE	VERTICAL	CHOKER	BASKET
1/4"	0.56	0.42	1.1
5/16"	0.87	0.66	1.7
3/8"	1.2	0.94	2.5
7/16"	1.7	1.3	3.4
1/2"	2.2	1.6	4.4
9/16"	2.7	2.1	5.5
5/8"	3.4	2.6	6.8
3/4"	4.8	3.7	9.7
7/8"	6.6	5	13
1"	8.3	6.4	17
1-1/8"	10	8.1	21
1-1/4"	13	9.9	26

All values listed are in U.S. tons

Rated loads based on:
1) a minimum D/d ratio of 25/1
2) pin diameter not less than the sling diameter

*NOTE: The capacity tables are only intended to assist the user by
providing typical sling data for job planning activities*

v5 r1- 5/17

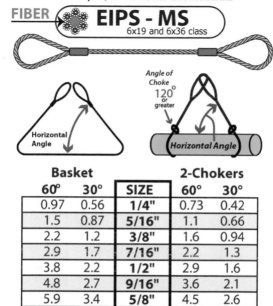

FIBER → EIPS - MS
6x19 and 6x36 class

Angle of Choke 120° **or greater**

Horizontal Angle

Basket		SIZE	2-Chokers	
60°	30°		60°	30°
0.97	0.56	1/4"	0.73	0.42
1.5	0.87	5/16"	1.1	0.66
2.2	1.2	3/8"	1.6	0.94
2.9	1.7	7/16"	2.2	1.3
3.8	2.2	1/2"	2.9	1.6
4.8	2.7	9/16"	3.6	2.1
5.9	3.4	5/8"	4.5	2.6
8.4	4.8	3/4"	6.3	3.7
11	6.6	7/8"	8.6	5
14	8.3	1"	11	6.4
18	10	1-1/8"	14	8.1
22	13	1-1/4"	17	9.9

Values listed in U.S. tons

Note: Rated loads based on a minimum D/d of 25:1

The values shown are listed in US Tons and are based on generic data
Always consult the sling manufacturer for the exact Working Load Limits

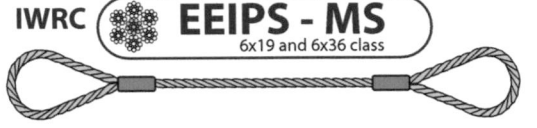

IWRC **EEIPS - MS**
6x19 and 6x36 class

Extra - EXTRA
Improved Plow,
Mechanical Splice

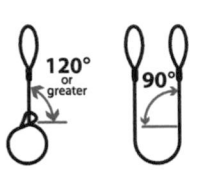

120° or greater

90°

SIZE	VERTICAL	CHOKER	BASKET
1/4"	0.71	0.52	1.4
5/16"	1.1	0.81	2.2
3/8"	1.6	1.2	3.2
7/16"	2.1	1.6	4.3
1/2"	2.8	2	5.5
9/16"	3.5	2.6	7
5/8"	4.3	3.2	8.6
3/4"	6.2	4.5	12
7/8"	8.3	6.1	17
1"	11	8	22

All values listed are in U.S. tons

Rated loads based on:
1) a minimum D/d ratio of 25/1
2) pin diameter not less than the sling diameter

Based on generic data - Always use the sling tag
to obtain the Working Load Limits (WLL)

*NOTE: The capacity tables are only intended to assist the user by
providing typical sling data for job planning activities*

v5 r1- 5/17

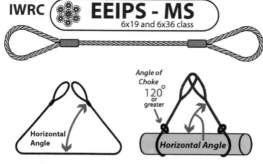

IWRC EEIPS - MS
6x19 and 6x36 class

Basket			2-Chokers	
60°	30°	*SIZE*	60°	30°
1.2	0.71	**1/4"**	0.9	0.52
1.9	1.1	**5/16"**	1.4	0.81
2.7	1.6	**3/8"**	2	1.2
3.7	2.1	**7/16"**	2.7	1.6
4.8	2.8	**1/2"**	3.5	2
6.1	3.5	**9/16"**	4.5	2.6
7.5	4.3	**5/8"**	5.5	3.2
11	6.2	**3/4"**	7.9	4.5
14	8.3	**7/8"**	11	6.1
19	11	**1"**	14	8

Values listed in U.S. tons

Note: Rated loads based on a minimum D/d of 25:1
    Basket data can also be applied to 2-leg bridles

The values shown are listed in US Tons and are based on generic data
Always consult the sling manufacturer for the exact Working Load Limits

3 - Leg Bridle

IWRC EIPS - MS 6x19 and 6x36 class

Extra Improved Plow Steel, Mechanical Splice

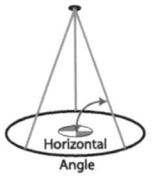

Horizontal Angle

SIZE	VERT	Horizontal Angle		
		60°	45°	30°
1/4"	1.9	1.7	1.4	0.97
5/16"	3	2.6	2.1	1.5
3/8"	4.3	3.7	3	2.2
7/16"	5.8	5	4.1	2.9
1/2"	7.6	6.6	5.4	3.8
9/16"	9.6	8.3	6.8	4.8
5/8"	12	10	8.3	5.9
3/4"	17	15	12	8.4
7/8"	23	20	16	11
1"	29	26	21	15
1-1/8"	36	31	26	18
1-1/4"	44	38	31	22
1-3/8"	53	46	38	27
1-1/2"	63	55	45	32
1-5/8"	73	63	52	37
1-3/4"	85	74	60	42
1-7/8"	97	84	68	48
2"	110	95	78	55

Rated loads for bridle slings are based on symmetrical loading with all legs being used

v5 r1- 5/17

4 - Leg Bridle

IWRC EIPS - MS 6x19 and 6x36 class

**Extra Improved
Plow Steel,
Mechanical Splice**

SIZE	VERT	Horizontal Angle		
		60°	45°	30°
1/4"	2.6	2.2	1.8	1.3
5/16"	4	3.5	2.8	2
3/8"	5.7	5	4.1	2.9
7/16"	7.8	6.7	5.5	3.9
1/2"	10	8.8	7.1	5.1
9/16"	13	11	9	6.4
5/8"	16	14	11	7.8
3/4"	22	19	16	11
7/8"	30	26	21	15
1"	39	34	28	20
1-1/8"	48	42	34	24
1-1/4"	59	51	42	30
1-3/8"	71	62	50	36
1-1/2"	84	73	60	42
1-5/8"	98	85	69	49
1-3/4"	113	98	80	57
1-7/8"	129	112	91	64
2"	147	127	104	73

Rated loads for bridle slings are based on symmetrical loading with all legs being used

WEB SLINGS EE
1-ply — Working Load Limits

LIGHT DUTY

EE1-6xx 1-ply, Class 5, EE light duty	Vertical	Choker (120° or greater)	90°	2-Leg or Basket 60°	45°	30°
1"	1,100	880	2,200	1,900	1,600	1,100
1 ½"	1,600	1,280	3,200	2,800	2,300	1,600
1 ¾"	1,900	1,520	3,800	3,300	2,700	1,900
2"	2,200	1,760	4,400	3,800	3,100	2,200
3"	3,300	2,640	6,600	5,700	4,700	3,300
4"	4,400	3,520	8,800	7,600	6,200	4,400
5"	5,500	4,400	11,000	9,500	7,800	5,500
6"	6,600	5,280	13,200	11,400	9,300	6,600

Based on generic data - Always use the sling tag to obtain the Working Load Limits (WLL)

2-ply WEB SLINGS EE
Working Load Limits

LIGHT DUTY

EE2-6xx 2-ply, Class 5, EE light duty	Vertical	Choker	90	60	45°	30
1"	2,200	1,760	4,400	3,800	3,100	2,200
1 1/2"	3,300	2,640	6,600	5,700	4,700	3,300
1 3/4"	3,800	3,040	7,600	6,600	5,400	3,800
2"	4,400	3,520	8,800	7,600	6,200	4,400
3"	6,600	5,280	13,200	11,400	9,300	6,600
4"	8,200	6,560	16,400	14,200	11,600	8,200
5"	10,200	8,160	20,400	17,700	14,400	10,200
6"	12,300	9,840	24,600	21,300	17,400	12,300

Based on generic data - Always use the sling tag to obtain the Working Load Limits (WLL)

-- Page 17 --

EE WEB SLINGS
Working Load Limits
1-ply

HEAVY DUTY

Based on generic data - Always use the sling tag to obtain the Working Load Limits (WLL)

EE1-9xx 1-ply, Class 7, EE Heavy duty	Vertical	Choker	90°	2-Leg or Basket		
				60°	45°	30°
1"	1,600	1,280	3,200	2,800	2,300	1,600
1 ½"	2,300	1,840	4,600	4,000	3,300	2,300
1 ¾"	2,700	2,160	5,400	4,700	3,800	2,700
2"	3,100	2,480	6,200	5,400	4,400	3,100
3"	4,700	3,760	9,400	8,100	6,600	4,700
4"	6,200	4,960	12,400	10,700	8,800	6,200
5"	7,800	6,240	15,600	13,500	11,000	7,800
6"	9,300	7,440	18,600	16,100	13,200	9,300
8"	11,750	9,400	21,150	18,300	15,000	11,750
10"	14,700	11,760	26,450	22,900	18,700	14,700
12"	17,650	14,120	31,750	27,500	22,400	17,650

2-ply WEB SLINGS EE
Working Load Limits

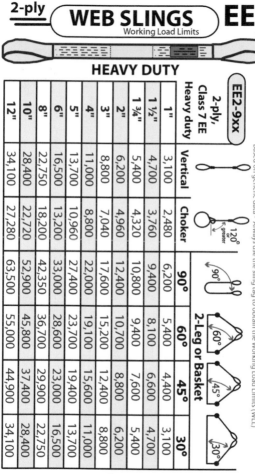

HEAVY DUTY

EE2-9xx

2-ply, Class 7 EE Heavy duty

Based on generic data - Always use the sling tag to obtain the Working Load Limits (WLL)

Heavy duty	Vertical	Choker	90°	60°	45°	30°
1"	3,100	2,480	6,200	5,400	4,400	3,100
1 ½"	4,700	3,760	9,400	8,100	6,600	4,700
1 ¾"	5,400	4,320	10,800	9,400	7,600	5,400
2"	6,200	4,960	12,400	10,700	8,800	6,200
3"	8,800	7,040	17,600	15,200	12,400	8,800
4"	11,000	8,800	22,000	19,100	15,600	11,000
5"	13,700	10,960	27,400	23,700	19,400	13,700
6"	16,500	13,200	33,000	28,600	23,000	16,500
8"	22,750	18,200	42,350	36,700	29,900	22,750
10"	28,400	22,720	52,900	45,800	37,400	28,400
12"	34,100	27,280	63,500	55,000	44,900	34,100

120° or greater

2-Leg or Basket

EN WEB SLINGS
1-ply Working Load Limits

LIGHT DUTY

Based on generic data - Always use the sling tag to obtain the Working Load Limits

EN1-60x

1-ply, Class 5, EN light duty	Vertical	Choker	90°	2-Leg or Basket		
				60°	45°	30°
1"	2,200	1,760	4,400	3,800	3,100	2,200
1 ½"	3,200	2,560	6,400	5,550	4,530	3,200
1 ¾"	3,800	3,040	7,600	6,600	5,400	3,800
2"	4,400	3,520	8,800	7,600	6,200	4,400
3"	6,600	5,280	13,200	11,400	9,300	6,600
4"	8,800	7,040	17,600	15,280	12,480	8,800
5"	11,000	8,800	22,000	19,100	15,600	11,000
6"	13,200	10,560	26,400	22,920	18,720	13,200

Values shown are in U.S. lbs

WEB SLINGS — EN

2-ply

Working Load Limits

LIGHT DUTY

EN2-60x

2-ply, Class 5, EN light duty	Vertical	Choker	2-Leg or Basket			
			90°	60°	45°	30°
1"	4,400	3,520	8,800	7,600	6,200	4,400
1 ½"	6,600	5,280	13,200	11,400	9,300	6,600
1 ¾"	7,600	6,080	15,200	13,000	10,000	7,600
2"	8,800	7,040	17,600	15,280	12,480	8,800
3"	13,200	10,560	26,400	22,920	18,720	13,200
4"	16,400	13,120	32,800	28,400	23,200	16,400
5"	20,400	16,320	40,800	35,400	28,900	20,400
6"	24,600	19,680	49,200	42,700	34,880	24,600

120° or greater

Based on generic data - Always use the sling tag to obtain the Working Load Limits

Values shown are in U.S. lbs

EN WEB SLINGS
1-ply · Working Load Limits

HEAVYDUTY

Based on generic data - Always use the sling tag to obtain the Working Load Limits

EN1-9xx
1-ply, Class 7, EN Heavy duty

	Vertical	Choker	2-Leg or Basket			
		(120°)	90°	60°	45°	30°
1"	3,200	2,560	6,400	5,550	4,530	3,200
1 ½"	4,600	3,680	9,200	7,987	6,524	4,600
1 ¾"	5,400	4,320	10,800	9,376	7,658	5,400
2"	6,200	4,960	12,400	19,100	15,600	11,000
3"	9,400	7,520	18,800	16,322	13,331	9,400
4"	12,400	9,920	24,800	21,531	17,585	12,400
5"	15,600	12,480	31,200	27,087	22,124	15,600
6"	18,600	14,880	37,200	32,296	26,378	18,600
8"	21,150	16,920	42,300	36,724	29,995	21,150
10"	26,450	21,160	52,900	45,927	37,511	26,450
12"	31,750	25,400	63,500	55,130	45,027	31,750

Values shown are in U.S. lbs

2-ply WEB SLINGS EN
Working Load Limits

EN2-9xx

Based on generic data - Always use the sling tag to obtain the Working Load Limits

2-ply, Class 7 EN Heavy duty	Vertical	Choker	2-Leg or Basket 90°	60°	45°	30°
1"	6,200	4,960	12,400	10,765	8,793	6,200
1 1/2"	9,400	7,520	18,800	16,322	13,331	9,400
1 3/4"	10,800	8,640	21,600	18,753	15,316	10,800
2"	12,400	9,920	24,800	21,531	17,585	12,400
3"	17,600	14,080	35,200	30,560	24,960	17,600
4"	22,000	17,600	44,000	38,200	31,200	22,000
5"	27,400	21,920	54,800	47,576	38,858	27,400
6"	33,000	26,400	66,000	57,300	46,800	33,000
8"	42,350	33,880	84,700	73,535	60,060	42,350
10"	52,900	42,320	105,800	91,854	75,022	52,900
12"	63,500	50,800	127,000	110,259	90,055	63,500

120° greater

Values shown are in U.S. lbs

ROUNDSLINGS
Working Load Limits

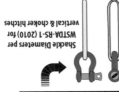

Based on generic sling data - Always use the sling tag to obtain the Working Load Limits

Size	Vertical	Choker	"Tight" Choke (see page 51)	Minimum Diameter Inch
1	2,600	2,100	1,029	7/16"
2	5,300	4,200	2,058	5/8"
3	8,400	6,700	3,283	3/4"
4	10,600	8,500	4,165	7/8"
5	13,200	10,600	5,194	1"
6	16,800	13,400	6,566	1-1/8"
7	21,200	17,000	8,330	1-3/16"
8	25,000	20,000	9,800	1-1/4"
9	31,000	24,800	12,152	1-1/2"
10	40,000	32,000	15,680	1-5/8"
11	53,000	42,400	20,776	2"
12	66,000	52,800	25,872	2-1/8"
13	90,000	72,000	35,280	2-1/2"

Always use roundslings by the rated capacity as indicated on the sling tag - never rely on the color! There are no requirements for sling colors in any standard.

Shackle Diameters per WSTDA-RS-1 (2010) for vertical & choker hitches

v5 r1–5/17

ROUNDSLINGS
Working Load Limits

EDGE Radius of load

Roundslings shall only be allowed to come into direct contact with edges if they are smooth and are well rounded to a suitable edge radius. The required size of the edge radius, depends on the sling capacity (as shown in the left column), and increases with the size of the sling. These values hold true regardless of the type of hitch.

* MIN Edge Radius	90°	60°	45°	30°	Minimum Shackle Diameter
		2-Leg or Basket			Inch
3/16"	5,200	4,500	3,700	2,600	9/16"
1/4"	10,600	9,200	7,500	5,300	7/8"
5/16"	16,800	14,500	11,900	8,400	1"-1/16"
5/16"	21,200	18,400	15,000	10,600	1-1/4"
3/8"	26,400	22,900	18,700	13,200	1-3/8"
7/16"	33,600	29,100	23,800	16,800	1-5/8"
7/16"	42,400	36,700	30,000	21,200	1-5/8"
7/16"	50,000	43,300	35,400	25,000	1-7/8"
1/2"	62,000	53,700	43,800	31,000	2"
9/16"	80,000	69,300	56,600	40,000	2-3/8"
11/16"	106,000	91,800	74,900	53,000	2-3/4"
3/4"	132,000	114,300	93,300	66,000	3"
7/8"	180,000	155,900	127,300	90,000	3-1/2"

*These values apply to roundslings that are fully tensioned to their rated capacity. When roundslings are tensioned to lower force values, the minimum radius values will be reduce accordingly. Refer to WSTDA RS-1 for specifics

CHAIN SLINGS
Working Load Limits

Alloy Steel Grade: **80**

Grade 80 Alloy Steel Chain Slings

SIZE	Vertical	Choker	2-Leg or Basket 60°	45°	30°
7/32" (5.5 mm)	2,100	1,700	3,600	3,000	2,100
9/32" (7 mm)	3,500	2,800	6,100	4,900	3,500
5/16" (8 mm)	4,500	3,600	7,800	6,400	4,500
3/8" (10 mm)	7,100	5,700	12,300	10,000	7,100
1/2" (13 mm)	12,000	9,600	20,800	17,000	12,000
5/8" (16 mm)	18,100	14,500	31,300	25,600	18,100
3/4" (20 mm)	28,300	22,600	49,000	40,000	28,300
7/8" (22 mm)	34,200	27,400	59,200	48,400	34,200
1" (26 mm)	47,700	38,200	82,600	67,400	47,700
1-1/4" (32 mm)	72,300	57,800	125,200	102,200	72,300

Values shown are in U.S. lbs

Based on generic data – Always use the sling tag to obtain the Working Load Limits

v5 r1 -5/17

CHAIN SLINGS
Working Load Limits

Alloy Steel
Grade: **80**

Steel Chain
Grade 80 Alloy Slings

3 or 4 Leg Bridle Sling
Double Basket Sling

SIZE	60°	45°	30°
7/32" (5.5 mm)	5,500	4,400	3,200
9/32" (7 mm)	9,100	7,400	5,200
5/16" (8 mm)	11,700	9,500	6,800
3/8" (10 mm)	18,400	15,100	10,600
1/2" (13 mm)	31,200	25,500	18,000
5/8" (16 mm)	47,000	38,400	27,100
3/4" (20 mm)	73,500	60,000	42,400
7/8" (22 mm)	88,900	72,500	51,300
1" (26 mm)	123,900	101,200	71,500
1-1/4" (32 mm)	187,800	153,400	108,400

Values shown are in U.S. lbs

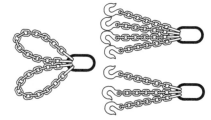

-- Page 27 --

CHAIN SLINGS

Grade: **100** Alloy Steel

Working Load Limits

Steel Chain Grade 100 Alloy Slings

SIZE		Vertical	Choker	2-Leg or Basket		
				60°	**45°**	**30°**
7/32"	(5.5 mm)	2,700	2,100	4,700	3,800	2,700
9/32"	(7 mm)	4,300	3,500	7,400	6,100	4,300
5/16"	(8 mm)	5,700	4,500	9,900	8,100	5,700
3/8"	(10 mm)	8,800	7,100	15,200	12,400	8,800
1/2"	(13 mm)	15,000	12,000	26,000	21,200	15,000
5/8"	(16 mm)	22,600	18,100	39,100	32,000	22,600
3/4"	(20 mm)	35,300	28,300	61,100	49,900	35,300
7/8"	(22 mm)	42,700	34,200	74,000	60,400	42,700

Values shown are in U.S. lbs

Based on generic data - Always use the sling tag to obtain the Working Load Limits

v5 r1-5/17

CHAIN SLINGS
Working Load Limits

Alloy Steel Grade: **100**

Steel Chain Grade 100 Alloy Slings

3 or 4 Leg Bridle Sling
Double Basket Sling

SIZE	60°	45°	30°
7/32" (5.5 mm)	7,000	5,700	4,000
9/32" (7 mm)	11,200	9,100	6,400
5/16" (8 mm)	14,800	12,100	8,500
3/8" (10 mm)	22,900	18,700	13,200
1/2" (13 mm)	39,000	31,800	22,500
5/8" (16 mm)	58,700	47,900	33,900
3/4" (20 mm)	91,700	74,900	53,000
7/8" (22 mm)	110,900	90,600	64,000

Based on generic data - Always use the sling tag to obtain the Working Load Limits

Values shown are in U.S. lbs

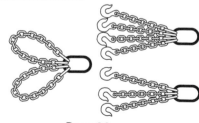

Eyebolts
Forged Alloy Machinery Eyebolts

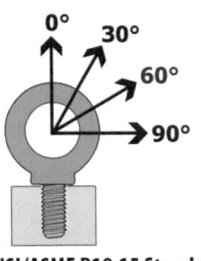

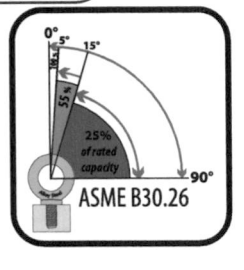

ASME B30.26

ANSI/ASME B18.15 Standard

SIZE	0°	30°	60°	90°
1/4"	400	75	NR	NR
5/16"	680	210	NR	NR
3/8"	1,000	400	220	180
7/16"	1,380	530	330	260
1/2"	1,840	850	520	440
9/16"	2,370	1,160	700	570
5/8"	2,940	1,410	890	740
3/4"	4,340	2,230	1,310	1,140
7/8"	6,000	2,960	1,910	1,630
1"	7,880	3,850	2,630	2,320
1-1/8"	9,920	4,790	3,840	3,390
1-1/4"	12,600	6,200	4,125	3,690
1-1/2"	18,260	9,010	6,040	5,460
1-3/4"	24,700	12,100	8,250	7,370
2"	32,500	15,970	10,910	9,740

WLL values shown in pounds

ANSI/ASME B18.15-1985 (Reaffirmed 1995, 2003)

ALWAYS VERIFY WLL's with the equipment manufacturer before use.

Eyebolts from an unknown manufacturers or with unknown material types should NEVER be used for overhead lifting!

v5 r1- 5/17

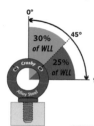

Eyebolts
Forged Alloy Machinery Eyebolts

If installing in tapped hole, make sure depth of thread engagement is at least 1.5 times bolt diameter. Where eyebolts must be aligned, a washer or shim may be placed under the shoulder to permit alignment when tightened. To minimize the bending moment, always apply load in the direction of the plane of the eye. Reduce working load limit according to tables if loaded other than true vertical.

Crosby®

S-279 Forged Machinery Eyebolts			
SIZE	0°	45°	90°
1/4" x 20	650	195	163
5/16" x 18	1,200	360	300
3/8" x 16	1,550	465	388
1/2" x 13	2,600	780	650
5/8" x 11	5,200	1,560	1,300
3/4" x 10	7,200	2,160	1,800
7/8" x 9	10,600	3,180	2,650
1" x 8	13,300	3,990	3,325
1-1/4" x 7	21,000	6,300	5,250
1-1/2" x 6	24,000	7,200	6,000

Forged Machinery Eyebolts

Chicago Hardware®			
#	SIZE	0°	45°
21	1/4" x 20	500	125
22	5/16" x 18	900	225
23	3/8" x 16	1,400	350
24	7/16" x 14	2,000	500
25	1/2" x 13	2,600	650
26	9/16" x 12	3,200	750
27	5/8" x 11	4,000	1,000
28	3/4" x 10	6,000	1,500
29	7/8" x 9	7,000	1,750
30	1" x 8	9,000	2,250
31	1-1/8" x 7	12,000	2,500
32	1-1/4" x 7	15,000	3,750
34	1-1/2" x 6	21,000	4,900

Chicago Hardware®

Shackles
Alloy - Screw Pin Anchor Type

Dimensions

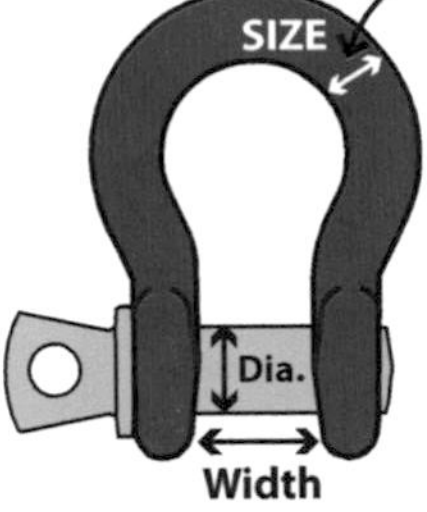

SIZE	Diameter of pin	Width
3/16"	0.25"	0.38"
1/4"	0.31"	0.47"
5/16"	0.38"	0.53"
3/8"	0.44"	0.66"
7/16"	0.50"	0.75"
1/2"	0.63"	0.81"
5/8"	0.75"	1.06"
3/4"	0.88"	1.25"
7/8"	1.00"	1.44"
1"	1.13"	1.69"
1-1/8"	1.25"	1.81"
1-1/4"	1.38"	2.03"
1-3/8"	1.5"	2.25"
1-1/2"	1.63"	2.38"
1-3/4"	2.00"	2.88"
2"	2.25"	3.25"
2-1/2"	2.75"	4.13"

The dimensions shown here were verified against Crosby and CM manufactured shackles. These are typical dimensions for most shackles, however they may vary with other manufacturers. All dimensions are in US inches and are approximate.

See page 46 for de-rating side loaded shackles

Shackles are designed and rated for inline applied tension. You can attach multiple slings in the body of a shackle, without reducing the capacity, provided that the shackle is symmetrically loaded AND the included angle does not exceed 120 degrees.

ASME B30·26

MAXIMUM Included Angle
120°

v5 r1- 5/17

Shackles

Shackle capacities will vary between manufacturers - this table is a summarized comparison that should be used for reference only. ALWAYS select and use the shackle by the rated capacity (WLL) that is shown on the shackle.

NEVER assume that all shackles are rated the same!

Forged ALLOY
Crosby® CM®

See note 3

SIZE	Carbon Steel See note 1 WLL tons	CM® Super Strong Carbon Steel See note 2 WLL tons	Forged ALLOY WLL tons	
3/16"	0.3	0.5		
1/4"	0.5	0.75		
5/16"	0.75	1		
3/8"	1	1.5	2	
7/16"	1.5	2	2.6	
1/2"	2	3	3.3	
5/8"	3.25	4.5	5	
3/4"	4.75	6.5	7	
7/8"	6.5	8.5	9.5	
1"	8.5	10	12.5	
1-1/8"	9.5	12	15	
1-1/4"	12	14	18	
1-3/8"	13.5	17	21	
1-1/2"	17	20	25 (SP)	30 (BT)
1-3/4"	25	30	34 (SP)	40 (BT)
2"	35	35	43 (SP)	55 (C-BT)
2-1/2"	55		85 (SP)	85 (C-BT)

note 1: Catalog data from Crosby, CM (Columbus McKinnon) and Chicago Hardware
note 2: Special type shackle available only from CM (Columbus McKinnon)
note 3: Catalog data from Crosby and CM (Columbus McKinnon)

(SP) = Screw Pin Type   (BT) = Bolt Type   (C-BT) = CROSBY Bolt Type

Anchor shackles manufactured by Crosby are rated in Metric tons - all others are rated in US tons

Swivel Hoist Rings

Same WLL regardless of the angle!

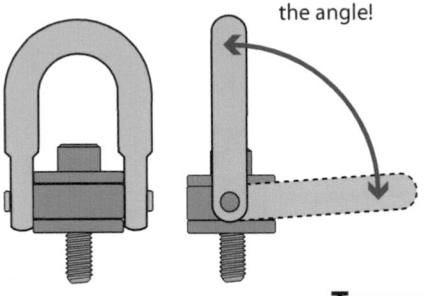

DIA x tpi *	WLL	Torque in ft lbs
5/16" x 18	800	7
3/8" x 16	1,000	12
1/2" x 13	2,500	28
5/8" x 11	4,000	60
3/4" x 10	5,000	100
7/8" x 9	8,000	160
1" x 8	10,000	230
1-1/4" x 7	15,000	470
1-1/2" x 6	24,000	800
2" x 4-1/2	30,000	1,100
2-1/2" x 4	50,000	2,100
3" x 4	75,000	4,300
3-1/2" x 4	100,000	6,600

* TPI (threads per inch)
The values shown are typical for most Swivel Hoist Rings, but may vary - alway verify with the manufacturer

v5 r1- 5/17

Wire Rope Clips
Installation data

Rope Diameter	No. of Clips	Turnback	Spacing	Torque in Foot-lbs (unlubed bolts)
1/8"	2	3-1/4"	3/4"	-
3/16"	2	3-3/4"	1-1/8"	-
1/4"	2	4-3/4"	1-1/2"	15
5/16"	2	5-1/2"	1-7/8"	30
3/8"	2	6-1/2"	2-1/4"	45
7/16"	2	7"	2-5/8"	65
1/2"	3	11-1/2"	3"	65
9/16"	3	12"	3-3/8"	95
5/8"	3	12"	3-3/4"	95
3/4"	4	18"	4-1/2"	130
7/8"	4	19"	5-1/4"	225
1"	5	26"	6"	225

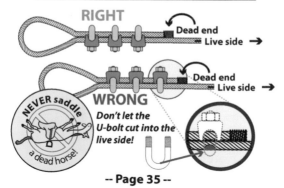

RIGHT

Dead end
Live side →

Dead end
Live side →

NEVER saddle a dead horse!

WRONG

Don't let the U-bolt cut into the live side!

Turnbuckles
Alloy Links

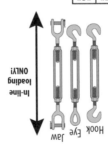

In-line loading ONLY!

Jaw Eye Hook

HOOK end fitting type	SIZE	EYE & JAW end fitting types
400	1/4"	500
700	5/16"	800
1,000	3/8"	1,200
1,500	1/2"	2,200
2,250	5/8"	3,500
3,000	3/4"	5,200
4,000	7/8"	7,200
5,000	1"	10,000
6,500	1-1/4"	15,200
7,500	1 1/2"	21,400

Based on Single leg (In-line load)

DIA "B"	WLL	"B"	"C"
1/2"	7,000	2.5	5
5/8"	9,000	3	6
3/4"	12,300	2.75	5.5
7/8"	15,000	3.75	6.38
1"	24,360	3.5	7
1-1/4"	36,200	4.38	8.75
1-1/2"	54,300	5.25	10.5
1-3/4"	84,900	6	12
2"	102,600	7	14
2-1/4"	143,100	8	16

Based on Single leg (In-line load)

DIA "E"	WLL	"D"	"E"
1/2"	7,000	2.5	5
5/8"	9,000	3	6
3/4"	12,300	2.75	5.5
7/8"	14,000	3.75	6.38
1"	24,360	3.5	7
1-1/4"	36,000	4.38	8.75
1-1/2"	54,300	5.25	10.5
1-3/4"	84,900	6	12
2"	102,600	7	14
2-1/4"	143,100	8	16

Obtained from 2006 Crosby Group catalog

Quick Sling Selection

Want a quick way to select the slings without calculating the tension? As long as you keep the horizontal sling angle above 60 degrees, select your slings as if EACH sling would support the entire weight of the load.

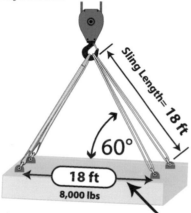

To have at least a 60 degree horizontal sling angle, measure between the attachment points, and select a sling that length (or longer)

Example:
Use 4 slings, EACH SLING with a Vertical capacity greater than 8,000 lbs

Graphics ©2012 Jerry Klinke

Tension Calc
using angles

Horizontal Angle	L.A.F.
5 °	11.49
10 °	5.75
15 °	3.861
20 °	2.924
25 °	2.364
30 °	2.00
35 °	1.742
40 °	1.555
45 °	1.414
50 °	1.305
55 °	1.221
60 °	1.155
65 °	1.104
70 °	1.064
75 °	1.035
80 °	1.015
85 °	1.004
90 °	1.00

To determine the amount of tension on a sling used at angles other than 90 degrees (vertical), use the table at the right to obtain the Load Angle Factor (L.A.F.) and following formula:

(Weight ÷ No of legs) x L.A.F.

Example: If the load weight is 4,000 lbs, and two (2) slings are used at a 40 degree angle each.

(4000 ÷2) x 1.555 = 3,110

Therefore, each leg will have 3,110 lbs of tension.

It is recommended that you consider that only 2 legs will carry the load, even when using 3 and 4 legs, since it is difficult to assure that all legs will carry an equal share of the load.

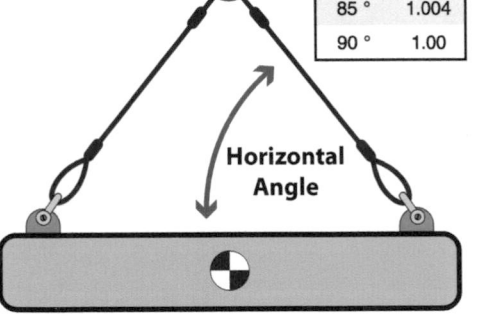

Horizontal Angle

v5 r1- 5/17

Tension Calc
using measurments

It's hard to determine the exact angle when working in the field unless you have a protractor handy. The following formula provides accurate calculations by using only measurements taken in the field:

(Weight ÷ No of legs) X (S ÷ H)

Example: The load weight is 6,000 lbs and two (2) slings are used. You measure up the sling 36" (this is the "S" dimension) then measure straight down and obtain a 24" measurment (this is the "H" dimension).

(6000 ÷ 2) x (36 ÷ 24) = 4,500 lbs of tension per leg

3000 x 1.5 = 4,500 lbs of tension per leg

It is recommended that you consider that only 2 legs will carry the load, even when using 3 and 4 legs, since it is difficult to assure that all legs will carry an equal share of the load.

You can also use the sling length and the vertical height

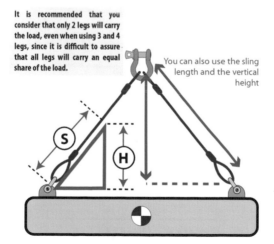

Drifting Loads

To determine how much tension will be placed upon chainfalls used in angular rigging situations, use the following formula:

Tension on Chainfall "A" =
(Load weight × D2 × LA) ÷ (H × D3)

Tension on Chainfall "B" =
(Load weight × D1 × LB) ÷ (H × D3)

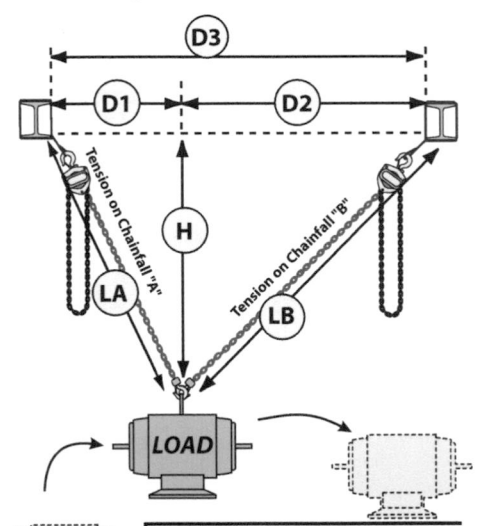

NOTE: This formula assumes that both chainfalls are positioned at the same elevation.

v5 r1- 5/17

Determine Load Share

You MUST know:
- ☑ the approximate CG location
- ☑ distance from CG to lift points
- ☑ approximate total weight

Load Share - LEFT end

$$(D2 / SPAN) \times TW = LEFT$$

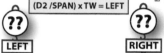

LEFT — ??

RIGHT — ??

D1   D2

SPAN

LEFT Lift Point

RIGHT Lift Point

Center of Gravity

Legend

CG = Center of Gravity
D1 = Distance from LEFT lift point to CG
D2 = Distance from RIGHT lift point to CG
SPAN = Distance between lift points
LEFT = Load share LEFT end
RIGHT = Load share RIGHT end
TW = Total Weight (approximate)

$$WL + WR = TW$$
$$WL / TW = \text{Percentage (of load share)}$$

Note: You only need to calculate one side, then subtract it from the TW for the other end

---

Find the Center of Gravity
by end weights

You MUST know the:
- ☑ weight of each end
- ☑ distance between lift points

Locate the CG from the LEFT lift point (D1)

$$(WR / TW) \times SPAN = D1$$

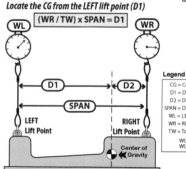

WL   WR

D1   D2

SPAN

LEFT Lift Point

RIGHT Lift Point

Center of Gravity

Legend

CG = Center of Gravity
D1 = Distance from LEFT lift point to CG
D2 = Distance from RIGHT lift point to CG
SPAN = Distance between lift points
WL = LEFT End Weight
WR = RIGHT End Weight
TW = Total Weight (WL+WR)

$$WL + WR = TW$$
$$WL / TW = \text{Percentage (of load share)}$$

Block Loading

A single sheave block used to change load line direction can be subjected to total loads greatly different from the weight being lifted or pulled. The total load value varies with the angle between the incoming and departing lines to the block.

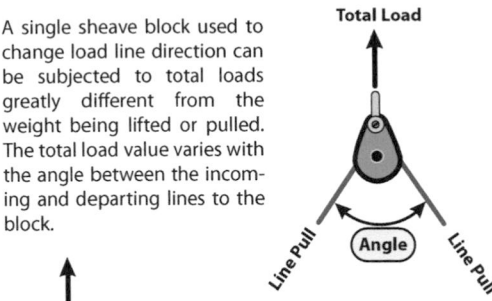

Total Load

Line Pull **Angle** Line Pull

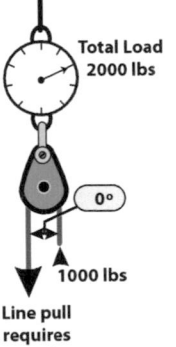

Total Load 2000 lbs

0°

1000 lbs

Line pull requires 1000 lbs

NOTE: The total load at the pulley attachment point is always greater than the weight of the load being lifted!

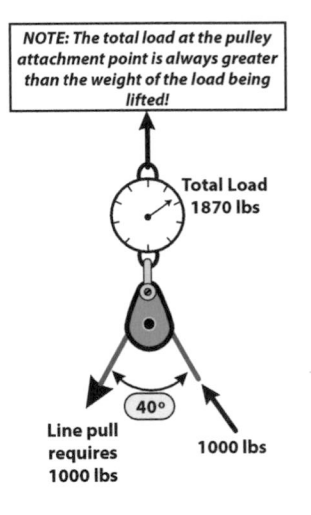

Total Load 1870 lbs

40°

Line pull requires 1000 lbs

1000 lbs

Total Load = Line Pull x Angle Factor

Example: 5000 lbs line pull at 40° (5000 x 1.87) a total load of 9,350 lbs

Angle	Factor	Angle	Factor
0°	2.00	100°	1.29
10°	1.99	110°	1.15
20°	1.97	120°	1.00
30°	1.93	130°	0.84
40°	1.87	135°	0.76
45°	1.84	140°	0.68
50°	1.81	150°	0.52
60°	1.73	160°	0.35
70°	1.64	170°	0.17
80°	1.53	180°	0
90°	1.41	--	--

Mechanical Advantage

$$\text{Pull Required} = \frac{\text{Weight to be lifted}}{\text{Mechanical Advantage}}$$

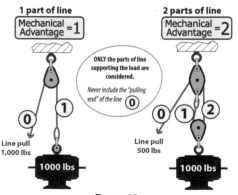

1 part of line

Mechanical Advantage = 1

ONLY the parts of line supporting the load are considered.

Never include the "pulling end" of the line ⓪

2 parts of line

Mechanical Advantage = 2

Line pull 1,000 lbs

1000 lbs

Line pull 500 lbs

1000 lbs

Pulling Force

To move a load
on a LEVEL plane

$$F = CF \times W$$

To move a load on an
UPHILL incline

$$F = [CF \times W \times (R \div L)] + [(H \div L) \times W]$$

To move a load on a DOWNHILL incline:

$$F = [CF \times W \times (R \div L)] - [(H \div L) \times W]$$

LEGEND	
W =	Weight of load
CF =	Coefficient of friction
F =	Force required to move load
H =	HEIGHT (Vertical distance in feet)
R =	RUN (Horizontal distance in feet)

These coefficient of friction values apply to hard, clean surfaces sliding against each other. These may not directly relate to your application due to actual surface conditions.

Materials	CF
Load on wheels/rollers	0.05
Steel on Steel (wet)	0.10
Metal on Wood (wet)	0.30
Wood on Wood (dry)	0.50
Wood on Concrete (dry)	0.50
Metal on Wood (dry)	0.60
Concrete on Concrete (dry)	0.65
Steel on Steel (dry)	0.74

Proper Sling Use

Never tie slings together !

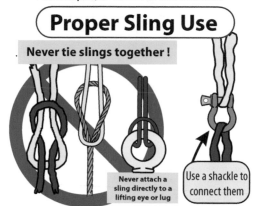

Never attach a sling directly to a lifting eye or lug

Use a shackle to connect them

Applies to all types of slings!

Sharp corners can cut and damage slings and result in sling failure.

NO

The sling can be permanently damaged!

WEAR PROTECTION

"Sharp edges in contact with the sling should be padded with material of sufficient strength to protect the sling."

ASME B30.9

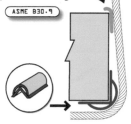

CORNER PROTECTION

Best practice: change the profile of a corner in contact with a sling to a radius.

Shackle Use

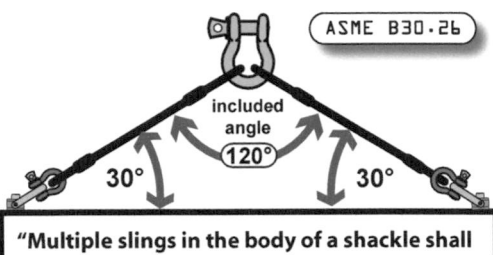

ASME B30.26

included angle 120°

30°   30°

> "Multiple slings in the body of a shackle shall not exceed 120 degree included angle"

NEVER side load Round Pin Shackles

If the shackle is to be side loaded **(with a SINGLE sling)** the rated load shall be reduced according to the recommendations of the manufacturer or a qualified person. ASME B30.26 recommends reducing the capacity of a side loaded shackle from 30% to 50% as shown below.

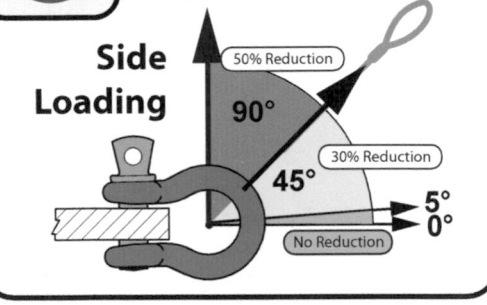

Side Loading

50% Reduction

90°

30% Reduction

45°

5°
0°

No Reduction

v5 r1- 5/17

Shackle Use

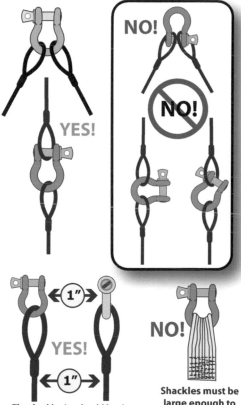

YES!

NO!

NO!

YES!

NO!

The shackle size should be the same diameter or larger than the wire rope size

Shackles must be large enough to avoid pinching of synthetic web slings

Eye Bolt Use

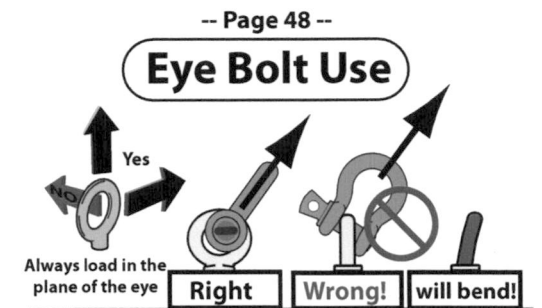

Always load in the plane of the eye

| **Right** | **Wrong!** | **will bend!** |

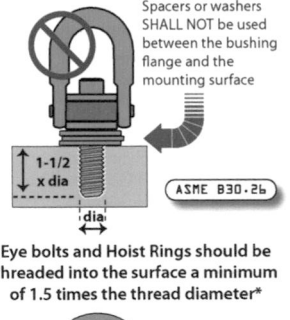

Spacers or washers SHALL NOT be used between the bushing flange and the mounting surface

ASME B30·26

1-1/2 x dia

dia

Eye bolts and Hoist Rings should be threaded into the surface a minimum of 1.5 times the thread diameter*

1-1/2" dia

1"

1/2"

1" dia

Flat washers may be used under the shoulder to position the plane of the eye

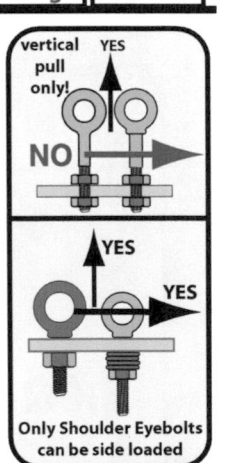

vertical pull only!

YES

NO

YES

YES

Only Shoulder Eyebolts can be side loaded

* per ASME B30.26-2.9.4.2 "when used in a tapped blind hole, the effective thread length shall be at least 1-1/2 times the diameter of the bolt for engagement in steel ... For other thread engagements or in other materials, contact the eyebolt manufacturer or a qualified person."

Hook Use

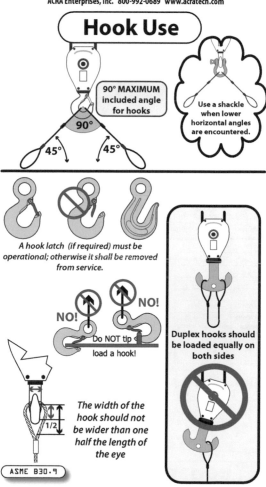

90° MAXIMUM included angle for hooks

90°

45°   45°

Use a shackle when lower horizontal angles are encountered.

A hook latch (if required) must be operational; otherwise it shall be removed from service.

NO!   NO!

Do NOT tip load a hook!

Duplex hooks should be loaded equally on both sides

1/2

The width of the hook should not be wider than one half the length of the eye

ASME B30.9

Choke Reduction

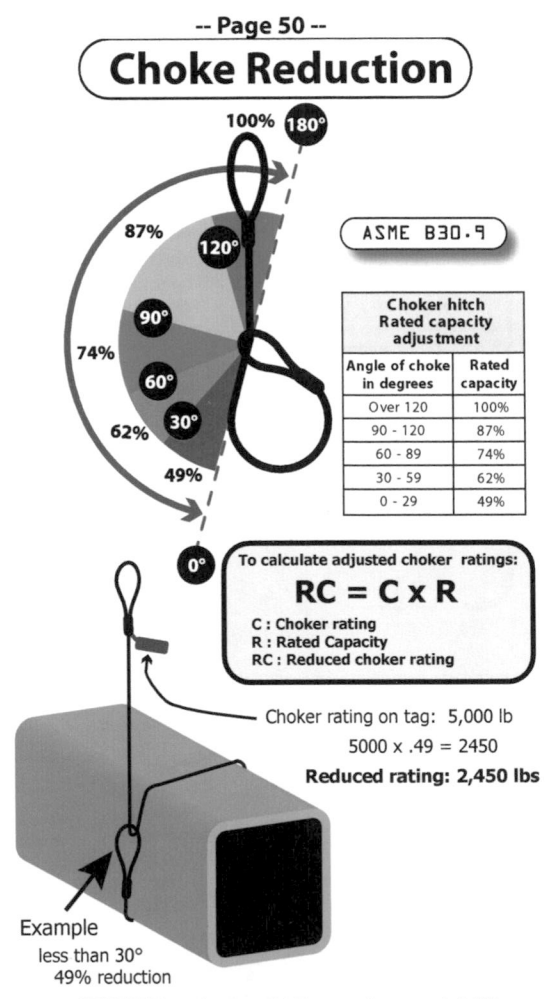

ASME B30.9

Choker hitch Rated capacity adjustment	
Angle of choke in degrees	Rated capacity
Over 120	100%
90 - 120	87%
60 - 89	74%
30 - 59	62%
0 - 29	49%

To calculate adjusted choker ratings:

RC = C x R

C : Choker rating
R : Rated Capacity
RC : Reduced choker rating

Choker rating on tag: 5,000 lb

5000 x .49 = 2450

Reduced rating: 2,450 lbs

Example
less than 30°
49% reduction

Minimum D/d for wire rope

Wire rope diameter MULTIPLY by 25 (or whatever the D/d ratio is)

WR Dia	25:1	WR Dia	25:1
1/4"	6-1/4"	7/8"	21-7/8"
5/16"	7-7/8"	1"	25"
3/8"	9-3/8"	1-1/8"	28-1/8"
1/2"	12-1/2"	1-1/4"	31-1/4"
5/8"	15-5/8"	1-3/8"	34-3/8"
3/4"	18-3/4"	1-1/2"	37-1/2"

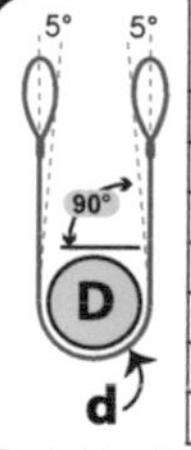

Table A

D/d	Efficiency
30	95%
20	92%
10	86%
5	75%
2	65%
1	50%

To calculate reduced capacity:

RC = B x E

For use with Wire Rope ONLY

6x19 or 6x36 (IWRC) Wire Rope
Mechanical: D/d ratio of 25/1
Hand-tucked: D/d ratio of 15/1
Braided Slings: D/d ratio of 25
  times the component rope diameter

B: Basket Rating
E: Efficiency (from table A)
M: Multiplier (from table B)
RC: Reduced Basket Rating
C: Choker Rating (listed on tag)

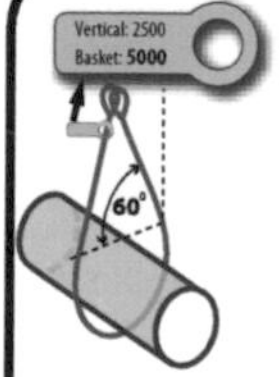

For use with ALL sling types

Table B

Vertical: 2500
Basket: 5000

Horizontal Angle	Multiplier
90°	1.000
60°	0.866
45°	0.707
30°	0.500

To calculate reduced capacity:

RC = B x M

OSHA Requirements

Key requirements from OSHA 1910.184

- Any damaged or defective sling shall not be used.
- Slings shall not be shortened with knots or bolts or other makeshift devices.
- Sling legs shall not be kinked.
- Slings shall not be loaded in excess of their rated capacities.
- Slings used in a basket hitch shall have the loads balanced to prevent slippage.
- Slings shall be securely attached to their loads.
- Slings shall be padded or protected from the sharp edges of their loads.
- Shock loading is prohibited.
- A sling shall not be pulled from under a load when the load is resting on the sling.
- Inspections: Each day before being used, the sling and all fastenings and attachments shall be inspected for damage or defects by a competent person designated by the employer.

Inspections (from ASME B30.9)

Initial Inspection; Before any new sling is placed into service it shall be inspected by a designated person.

Frequent Inspection; An inspection shall be performed by the user or other designated person each day or shift the sling is used.

Periodic Inspection; A complete inspection for damage of the sling shall be periodically performed by a designated person.

Record keeping and documentation may be required (consult ASME B30.9 for specifics)

OSHA Requirements

Slings shall be immediately removed from service if any of the following conditions are present:

Synthetic web slings:

- Acid or caustic burns
- Melting or charring of any part of the sling surface
- Snags, punctures, tears or cuts
- Broken or worn stitches or Distortion of fittings
- Missing or illegible identification tag

Wire rope slings:

- 10 randomly distributed broken wires in 1 rope lay, or 5 broken wires in 1 strand in 1 rope lay
- Wear or scraping of one-third the original diameter of outside individual wires
- Kinking, crushing, bird caging or any other damage resulting in distortion of the wire rope structure
- Evidence of heat damage
- Missing or illegible identification tag

Roundslings

- Acid or caustic burns; heat damage; melting
- Holes, tears, cuts, or snags that expose the core yarns
- Broken or damaged core yarns; Knots in the sling
- Discoloration, brittle or stiff areas
- Missing or illegible identification tag

Alloy Chain slings

- Cracks or breaks, wear, nicks, gouges
- Stretched chain links, bent, twisted or deformed links
- Evidence of heat damage, weld splatter, pitting or corrosion
- Lack of ability to hinge freely
- Excessive wear on any point on a link
- Other conditions that cause doubt as to the continued use of the sling
- Missing or illegible identification tag

Does not cover all requirements - Consult OSHA 1910.184 and ASME B30.9

Reference

To find an unknown side of a right angle triangle:

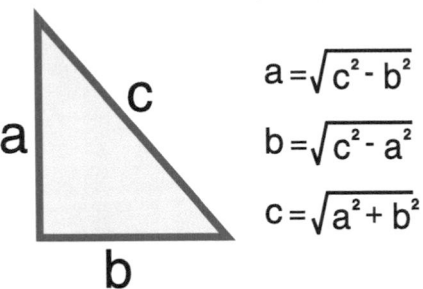

$$a = \sqrt{c^2 - b^2}$$

$$b = \sqrt{c^2 - a^2}$$

$$c = \sqrt{a^2 + b^2}$$

UNITS OF MEASURE

1 US ton (short)	=	2000 lbs
1 US ton (short)	=	.91 metric ton
1 US ton (short)	=	907 kgs
1 metric ton	=	2204.62 lbs
1 metric ton	=	1.102 US tons
1 metric ton	=	1000 kgs
1 US pound (lb)	=	16 ounces
1 US pound (lb)	=	.45 kg
1 kilogram (kg)	=	2.2 lbs
1 kilogram (kg)	=	35 ounces
1 kilogram (kg)	=	1000 grams
1 US (liq) gallon	=	4 quarts
1 US (liq) gallon	=	3.8 liters
1 liter	=	.264 gallons (US)
1 liter	=	1.06 quarts
1 US gallon water	=	8.3 lbs
1 cubic ft of liquid	=	7.5 US gallons

Reference

Standard weights of typical materials

Material	Cu. ft	Cu. Inch
Aluminum	165.00	0.0955
Brass	535.00	0.3096
Brick masonry, common	125.00	0 0723
Bronze	500.00	0.2894
Cast Iron	480.00	0.2778
Cement, portland, loose	94.00	0.0544
Concrete, stone aggr.	144.00	0.0833
Copper	560.00	0.3241
Earth, dry	75.00	0.0434
Earth, wet	100.00	0.0579
Ice	56.00	0.0324
Lead	710.00	0.4109
Snow, fresh fallen	8.00	0.0046
Snow, wet	35.00	0.0203
Steel	490.00	0.2836
Tin	460.00	0.2662
Water	62.00	0.0359
Gypsum wall board	54.00	0.0313
Wood, pine	30.00	0.0174

WEIGHT = VOLUME X WEIGHT per Cu...

Cube or Rectangle:
VOLUME = a X b X c

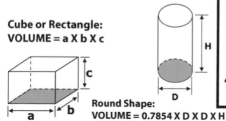

Round Shape:
VOLUME = 0.7854 X D X D X H

Mobile Crane

Hand Signals

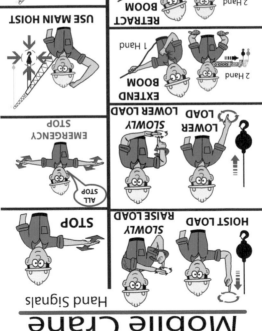

HOIST LOAD / **RAISE LOAD SLOWLY**

STOP

LOWER LOAD / **LOWER LOAD SLOWLY**

EMERGENCY STOP
ALL STOP

EXTEND BOOM
2 Hand / 1 Hand

RETRACT BOOM
1 Hand / 2 Hand

USE MAIN HOIST

SWING BOOM
DOG EVERYTHING

USE WHIP LINE

Refer to ASME B30.5 and OSHA 1926 for additional information

v5 r1-5/17

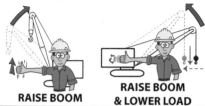

RAISE BOOM

RAISE BOOM & LOWER LOAD

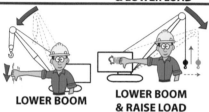

LOWER BOOM

LOWER BOOM & RAISE LOAD

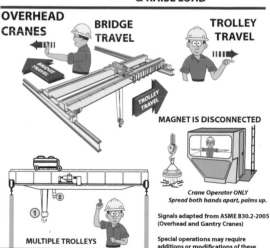

OVERHEAD CRANES

BRIDGE TRAVEL

TROLLEY TRAVEL

MAGNET IS DISCONNECTED

Crane Operator ONLY
Spread both hands apart, palms up.

MULTIPLE TROLLEYS

Hold up ONE finger for the block marked "1" and TWO fingers for the block Marked "2". Then follow with regular hand signals.

Signals adapted from ASME B30.2-2005 (Overhead and Gantry Cranes)

Special operations may require additions or modifications of these signals.

Signal person *(brief summary of 2010 OSHA rule)*
refer to OSHA 29CFR 1926.1428 for specific details

The signal person must:
- have a basic understanding of crane operations, limitations, crane dynamics, and boom deflection
- have demonstrated through a written and a practical test, understanding of the methods for crane hand signals
- have documentation from a qualified evaluator, or a third party, that they meet the requirements

Hand Signals - see pages 56/57
Voice Signals - must contain the following 3 elements, given in the following order:
   1. function and direction:
   2. distance and/or speed
   3. **function - stop command**
Examples....
*"Swing right: 30 feet, 20 feet, 10 feet, **Swing stop**"*
*"Hoist down: slow, slow, slow, **Hoist Stop**"*

- All directions are from Crane Operator's perspective
- Before the lift, the operator, signal person and lift director, must meet and agree on the voice signals to be used

v5 r1- 5/17

Phonetic Alphabet

Letter	Pronunciation
A	Alpha (AL fah)
B	Bravo (BRAH VOH)
C	Charlie (CHAR lee)
D	Delta (DELL tah)
E	Echo (ECK oh)
F	Foxtrot (FOKS trot)
G	Golf (GOLF)
H	Hotel (hoh TELL)
I	India (IN dee ah)
J	Juliett (JEW lee ETT)
K	Kilo (KEY loh)
L	Lima (LEE mah)
M	Mike (MIKE)
N	November (no VEM ber)
O	Oscar (OSS cah)
P	Papa (pah PAH)
Q	Quebec (keh BECK)
R	Romeo (ROW me oh)
S	Sierra (see AIR rah)
T	Tango (TANG go)
U	Uniform (YOU nee form)
V	Victor (VIK tah)
W	Whiskey (WISS key)
X	X Ray (ECKS RAY)
Y	Yankee (YANG key)
Z	Zulu (ZOO loo)

Note: The syllables printed in capital letters are to be stressed

Protecting Slings

There 2 different protection types:

ABRASION PROTECTION

to protect the sling from abrasion against the load surface

CUT PROTECTION

to protect the sling from being cut by an edge or corner of the load

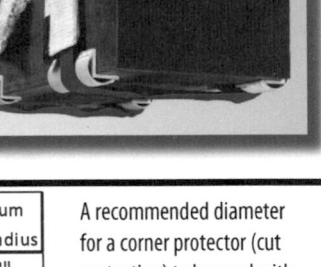

WIRE ROPE DIA	Minimum corner radius
1/4"	1-3/4"
5/16"	2-1/8"
3/8"	2-5/8"
1/2"	3-1/2"
5/8"	4-3/8"
3/4"	5-1/4"
7/8"	6-1/8"
1"	7"
1-1/8"	7-7/8"
1-1/4"	8-3/4"

A recommended diameter for a corner protector (cut protection) to be used with wire rope is approximately 7 times the diameter* of the wire rope, as shown in this example.

1/2" wire rope

7" diameter pipe
(cut in half) ➜

* This minimum diameter guideline is considered a good practice and followed by many companies.

v5 r1- 5/17